PERIPLUS NATURE GUIDES

TROPICAL
BIRDS
of Malaysia & Singapore

Text and photographs by Morten Strange

PERIPLUS

EDITIONS

Published by Periplus Editions (HK) Ltd.

Copyright © 1998 Periplus Editions (HK) Ltd.
ALL RIGHTS RESERVED
Printed in the Republic of Singapore
ISBN 962-593-167-8

Publisher: Eric M. Oey
Design: Peter Ivey
Editor: Julia Henderson
Production: Agnes Tan
Cover photographs: Morten Strange
Additional photographs by Lassie Laine,
Pete Morris, Ong Kiem Sian, Atsuo Tsuji,
Eric Woods and Alan Ow Yong

Distributors
Indonesia
PT Wira Mandala Pustaka
(Java Books-Indonesia)
Jalan Kelapa Gading Kirana,
Blok A14 No. 17,
Jakarta 14240

Singapore and Malaysia
Berkeley Books Pte. Ltd.,
5 Little Road #08-01, Singapore 536983

United States
Charles E. Tuttle Co., Inc.,
RRI Box 231-5, North Clarendon,
VT 05759-9700

Introduction

Southeast Asia is home to a remarkable array of bird species—the most diverse on Earth. These birds range in size from the tiny Orange-bellied Flowerpeckers living deep within the rainforest, to massive White-bellied Sea-Eagles soaring high above the open seas. They vary in color from drab, olive babblers, to exquisitely-hued kingfishers. Lively sandpipers run on the open mudflats, while inert, enigmatic trogons perch motionless just below the rainforest canopy.

Three well-defined, zoogeographical realms lie within Southeast Asia: the Oriental region (extending from Pakistan across southern China, Taiwan, the Philippines and into Indonesia); the Australasian region (reaching from the island of New Guinea, south-east across Australia including New Zealand); and a transitional subregion, known as *Wallacea*, located between the Oriental and the Australasian regions (covering central Indonesia). Taxonomic research carried out in 1996 by the Oriental Bird Club showed that the Oriental region alone was found to contain a staggering 2,586 bird species. Some are endemic to specific areas or single islands, especially in the Philippines and Indonesia, while others are widespread and found in most countries.

Residents and visitors alike are taking an increasing interest in the birdlife around them, however birdwatching (or "birding") is not that easy in Southeast Asia. Evergreen tropical rainforests are usually the natural vegetation cover and inside these thickly-canopied forests the birds are usually difficult to find and see clearly. These forest birds occupy narrow niches in a complex and static ecosystem—they reproduce slowly and the nests of many have never been found. An alarming number are in danger of extinction. Consequently, national parks and nature reserves have been created throughout the region to protect bird habitats that are, in many cases, disappearing.

The majority of species native to Southeast Asia are unable to adapt to life in villages and gardens—those that do adjust are mostly coastal mangrove residents. A mature garden can be visited by up to fifty of these species during a year and more can be attracted if the right vegetation or a feeding table with fruits and meal worms is provided. The birds we have selected for this Nature Guide are mainly the easy-to-find species that have adapted to living in disturbed habitats near human habitation. These familiar birds are interspersed with some of the more specialized individuals found only in forest habitats. All are presented in conventional taxonomic order.

It is our hope that this introductory guide will help you enjoy the birdlife found in your everyday surroundings and in Southeast Asia's many excellent nature reserves.

Morten Strange

Little Heron

Butorides striatus

Family:
Ardeidae

Distribution:
Tropics
worldwide;
Thailand,
Indonesia,
Malaysia, the
Philippines and
Singapore.

The most widespread and common member of the Ardeidae family, the Little Heron is resident throughout Southeast Asia. Like all other family members, it has long legs and a strong beak used for catching fish and other aquatic prey. This heron can be found on coastal mudflats and beaches as well as near reservoirs and rivers far inland, usually seen as solitary individuals or in pairs. However during the northern winter individuals from the subtropical regions migrate south and this species then becomes more numerous, forming loose flocks around the coastlines.

The Little Heron stands near the water's edge, quietly waiting for small fishes to swim near, then suddenly rushes forward to capture its meal. When disturbed it flies off at a low angle, with strong wingbeats, crying out with a harsh abrupt call. The nest is a collection of sticks built low in a tree or in a bush near wetlands. In a suitable habitat several pairs will nest close to each other.

Javan Pond Heron

Ardeola speciosa

Two pond herons, both members of the *Ardeola* genus, are found in Southeast Asia. The Javan Pond Heron, *A. speciosa*, is a resident species and the Chinese Pond Heron, *A. bacchus*, is a winter visitor. In winter plumage, the two species have a similiar appearance.

Family:
Ardeidae

Distribution:
Southeast Asia;
Indonesia,
Malaysia and
Thailand.

The Javan Pond Heron can be very common on wetlands on Java and Bali, especially freshwater marshes and rice fields in the lowlands, and is also often found around fish ponds and brackish estuaries near the coast. You may find it difficult to get close to this attractive bird, since it is fairly shy. The photograph on the right was taken from a hide.

During the breeding season, in the first months of the year, this heron develops a handsome set of dark, chestnut-colored plumes that cover its back and breast, as shown in the photograph. The wings however remain pure white and flash brightly when the bird takes flight. The Javan Pond Heron stalks insects and small fish near the water's edge, as do most other herons, and sometimes a few are seen feeding together.

The Javan Pond Heron breeds only in Indonesia, Thailand and parts of Indochina. It builds its nest high in a tree, usually in colonies along with white egrets or other species of heron.

Egrets

Family:
Ardeidae

Distributions:

Little Egret and
Plumed Egret:
South and East
Asia; Indonesia,
Malaysia,
Thailand, the
Philippines, and
Singapore.

Great Egret:
Tropics world-
wide; Indonesia,
Malaysia, the
Philippines,
Thailand and
Singapore.

Pacific Reef-Egret:
East Asia and
Australia;
Indonesia,
Malaysia,
Thailand, the
Philippines and
Singapore.

Chinese Egret:
East Asia;
Indonesia,
Malaysia,
Thailand, the
Philippines and
Singapore.

Cattle Egret:
Tropics world-
wide; Indonesia,
Malaysia,
Thailand, the
Philippines and
Singapore.

Egrets are white members of the heron family. All are active, attractive wetland birds that nest in colonies in tall trees or on coastal cliffs. Outside the breeding season they are often seen along estuaries and shallow coastlines. Most species are of the *Egretta* genus. No less than six species occur in Southeast Asia and since they can only be recognized by color, body size, shape and soft parts, great care should be taken during identification.

Little Egret, *Egretta garzetta*: Common, small size, with a thin, black bill; usually found fishing in the shallow water along coastlines and flooded fields (top left).

Plumed Egret, *Egretta intermedia*: Of medium size, with a short, yellow bill. The Plumed Egret is found mostly in freshwater swamps, along estuaries and on tidal mudflats and is only common locally (top right).

Great Egret, *Egretta alba*: The largest of the egrets, the Great Egret is fairly common in all kinds of extensive wetlands. It is a slow-moving bird with a long neck and a huge bill that is yellow in winter and black during the breeding season (center left).

Pacific Reef-Egret, *Egretta sacra*: The combination of small size and strong pale bill is diagnostic. In northern Asia a dark morph is common. The Pacific Reef-Egret is always found near the coast, often along sandy beaches and on offshore rocky islands. It can be seen rushing through the water at low tide catching fish (center right).

Chinese Egret, *Egretta eulophotes*: Similar in shape to the Reef-Egret. Difficult to identify during winter, but in summer plumage its long crest and bright yellow bill is diagnostic. A rare coastal migrant from China and Korea, it is considered an endangered species (bottom left).

Cattle Egret, *Bubulcus ibis*: The smallest egret, with a stocky build and yellow bill. When in breeding plumage, the head and neck are orange. Locally very common, it is found in flocks near marshes and open fields, sometimes far from water and never at the coast. (bottom right).

Lesser Adjutant

Leptoptilos javanicus

Family:
Ciconidae

Distribution:
South Asia;
Indonesia,
Malaysia and
Thailand.

Storks, physically similar to herons, are larger water birds with long legs and beaks. Storks are easily recognized in flight as they hold their necks straight, while the heron bends its neck back towards its body. However, storks have become rare in Southeast Asia.

The Lesser Adjutant is a widespread resident throughout the region but not numerous anywhere. In Singapore this stork has become locally extinct and the species is considered vulnerable to global extinction elsewhere.

The Lesser Adjutant belongs to the same genus as the Marabou Stork, *L. crumeniferus*, of Africa and is somewhat similar to it in appearance. While the Marabou stork is a common scavenger and sometimes feeds on rubbish dumps near settlements, the Lesser Adjutant is a shy bird found only along remote seashores and around mangrove forests. A good pair of binoculars is necessary if you want to see the Lesser Adjutant feeding, however these huge birds are easy to spot and identify once in the air. At low tide the storks move out from the mangrove forests to feed on the exposed mudflats, taking mudskippers and other fish as well as crustaceans and frogs. They also feed occasionally on small reptiles and mammals found near the shore.

The Lesser Adjutant builds a large nest of sticks in an isolated and inaccessible patch of mangrove forest. Within a prime habitat several pairs will form small colonies, often together with other storks and herons.

Lesser Tree-Duck

Dendrocygna javanica

The natural vegetation of Southeast Asia is often characterized by dense rainforests growing from the hilltops down to the seashore. The few lakes and freshwater swamps in this part of the tropics are not particularly productive, thus the family of ducks, geese and swans (Anatidae) is poorly represented here—in fact this duck is the only really common resident of the region. Tree-ducks (also known as whistling-ducks because of their thin, whining calls) are members of the genus *Dendrocygna*.

The Lesser Tree-Duck, which feeds almost entirely on the seeds and shoots of water plants and other vegetable matter, is found around freshwater swamps, artificial reservoirs and along rivers, where plenty of weed and reeds grow along their banks. The female builds its nest in a bed of tall reeds near the edge of the water, in a hollow log or sometimes takes over an abandoned heron's nest. The duck is not particular about where it nests, as long as the fledglings can reach the water immediately after hatching.

Family:
Anatidae

Distribution:
South Asia;
Indonesia,
Malaysia,
Thailand and
Singapore.

Crested Serpent-Eagle

Spilornis cheela

Family:
Accipitridae

Distribution:
South Asia;
Indonesia,
Malaysia,
Thailand, the
Philippines and
Singapore.

Probably the most widespread and common of the birds of prey, this member of the hawk family is found all over Southeast Asia. Many subspecies and some closely-related species occur on offshore islands throughout its range. The Crested Serpent-Eagle is essentially a forest bird that lives in both primary rainforest and logged secondary forest and mangroves. It can often be spotted perched high on an open branch near the edge of a road or a clearing. After ten minutes to half an hour in this position, it will suddenly fly down and snatch its prey from the ground below—usually snakes but sometimes lizards and small mammals.

Late in the morning this raptor will often use the heat thermals to soar high over the forest. It marks its territory this way and signals to other eagles with a sharp, penetrating call that carries a long way. In flight this call plus a broad white band in the flight feathers and tail make the Crested Serpent-Eagle unmistakable.

Brahminy Kite

Haliastur indus

This pretty and elegant bird of prey is very common throughout Southeast Asia. An open country species, it is often numerous near the coast but can also be found upcountry in cultivated areas near rivers and along forest edges. An adult is shown below. The immature bird (under 3 years of age) is uniformly brown, with mottled white on the head and chest.

The Brahminy Kite lives on fish and meats. Although capable of lifting live fish out of the water, most of the time it scavenges on the beach or picks up edible debris floating on the water's surface. Away from the coast its diet includes insects, reptiles and small mammals. The Brahminy Kite is attracted to harbors and prawn farms where often several can be found feeding acrobatically quite close to humans, making this bird easy to find and exciting to watch. In between feeding sorties it rests on branches near the water. The nest is built in a large tree in a mangrove forest or coastal woodlands. Both parents attend to the young.

Family:
Accipitridae

Distribution:
South Asia, Australia; Indonesia, Malaysia, Thailand, the Philippines and Singapore.

White-bellied Sea-Eagle

Haliaeetus leucogaster

Family:
Accipitridae

Distribution:
South Asia;
Australia,
Indonesia,
Malaysia,
Thailand, the
Philippines and
Singapore.

This massive bird of prey is fortunately still a common sight in remote coastal areas of Southeast Asia, although it seems to be declining in numbers. It frequents coastal habitats—both forested mangroves, mudflats and exposed beaches with rocky offshore outcrops. This sea eagle is also sometimes found near the coast along large rivers, lakes and reservoirs. Each resident pair requires an extensive territory and although conspicuous, this eagle is not numerous anywhere in the region. The adult has a white head and underparts with grey wings, however this plumage does not develop until the bird is 5–6 years old. The immature bird, as shown on the left, is a mottled brown.

Like most tropical resident birds, the sea eagles are sedentary and do not migrate. They can often be seen late in the morning, soaring high above their territory, building up height over the land, using the coastal heat thermals for updrift, then shifting out over the water. While soaring they hold their wings motionless in a lifted V-angle, unlike most other birds of prey which hold their wings straight out. The White-bellied Sea-Eagle often catches live fish and is a spectacular sight as it swoops down, talons extended, to lift a large fish from the water's surface without wetting a feather. It also feeds on offal and scavenges on dead fish at the beach.

During the breeding season in the beginning of the year the pair performs acrobatic aerial displays, tumbling together through the air and crying out with loud goose-like honks. The nest is a huge construction of dead branches built on a remote seashore cliff or in a tall tree. In Malaysia and Singapore this sea eagle has been known to nest high in steel communication towers. The same nest is often used year after year.

Black-shouldered Kite

Elanus caeruleus

This handsome raptor belongs to the Accipitridae family of hawks and eagles, but appears almost falcon-like in the air. It has narrow, pointed wings and often hovers low above the ground before dropping onto prey in the grass below. This kite is unmistakable, with its small size and grey and black appearance. An open country species found throughout Southeast Asia, it is often seen in open woodlands, along coastal forest edges and over grasslands with scattered trees. It quickly turns up near coastal construction sites on reclaimed land.

A creature of habit, the Black-shouldered Kite stakes out a certain territory and often uses the same branch or pole for a perch day after day. From this perch it makes low flights across the terrain looking for small mammals, reptiles and large insects. A mating pair will perform splendid aerial displays during the beginning of the breeding season. The nest is built of sticks in a large tree near open country.

Family:
Accipitridae

Distribution:
South Asia;
Indonesia,
Malaysia,
Thailand,
the Philippines
and Singapore.

Pheasants

Family:
Phasianidae

Top left: Crested Wood-Partridge, *Rollulus rouloul*, occurs in Malaysia, south Thailand, Borneo and Sumatra.

Top right: Ferruginous Wood-Partridge, *Caloperdix oculea*, occurs in south Thailand, Malaysia, Borneo and Sumatra.

Center left: Bronze-tailed Peacock-Pheasant, *Polyplectron chalcurum*, occurs in Sumatra.

Center right: Great Argus, *Argusianus argus*, occurs in south Thailand, Malaysia, Borneo and Sumatra.

Bottom left: Red Junglefowl, *Gallus gallus*, occurs throughout South Asia, Indonesia, Malaysia, the Philippines, Singapore and Thailand.

Bottom right: Palawan Peacock-Pheasant, *Polyplectron emphanum*, occurs on the island of Palawan, in the Philippines.

The pheasants are a diverse family of birds found throughout Southeast Asia. Species range in size from the 245-cm male Green Peafowl, *Pavo muticus*, to the 15-cm Blue-breasted Quail, *Coturnix chinensis*. All pheasants and partridges live on the ground and prefer to run away from danger rather than fly. When forced to fly they take off at a low angle, with strong wing beats. They all have solid legs and rather plump bodies, and walk along the ground picking up seeds, small fruits and insects. All species nest on the ground and often have large broods. The young are capable of walking and finding food themselves almost immediately after hatching. Many males are very colorful with long ornamental tail feathers that they exhibit to the female on certain display grounds.

Pheasants are notoriously difficult to observe in the wild. Apart from the genus of quails, all species of pheasant in Southeast Asia are forest birds and prefer primary (undisturbed) rainforest. Comparative surveys have shown that this family of birds is poorly equipped to adapt to a secondary habitat, presumably because the forest undergrowth becomes too dense once the large trees are logged out and more light reaches the forest floor. As a result most species are rare and can only be found in remote parts of national parks and forest reserves. Many are endemic to specific areas and some are restricted to montane forest habitat. Furthermore pheasants are extremely wary and shy and will usually wander off long before it is possible to see them—a brief glimpse as one crosses the forest trail is often the only view possible.

The exception to this rule is the Red Junglefowl, *Gallus gallus*, the ancestor of the domestic chicken, which does well in secondary forest and along forest edges. This bird is often seen and heard in Southeast Asia. Calls are a good way of locating the elusive pheasants—the booming call of the Great Argus, *Argusianus argus*, provides an exotic ingredient to a morning walk in the rainforest.

White-breasted Waterhen

Amaurornis phoenicurus

Family:
Rallidae

Distribution:
South Asia;
Indonesia, the
Philippines,
Malaysia,
Thailand and
Singapore.

This delightful member of the rail family, common throughout Southeast Asia, is found near freshwater swamps, lakes and rivers and sometimes around coastal mangroves. Unlike other rails, the White-breasted Waterhen will come out into the open near rice fields and even into gardens and parks as long as water is nearby. The waterhen is often seen running across vegetation on the water's surface, searching for aquatic invertebrate prey.

Local people in the region refer to this fairly conspicuous bird as the "chicken bird," however it is in no way related to the family of pheasants. Although it is often found living near humans, the waterhen never becomes really tame and will run into cover when disturbed. It only flies a short distance at a time—just far enough to dive into the nearest tall grass. The adult White-breasted Waterhen is quite vocal and especially during dusk it will call out loudly from its hiding place with a peculiar series of croaks and chuckles.

Ruddy-breasted Crake

Porzana fusca

Rails are secretive birds with strong legs usually found living near freshwater marshes. The Ruddy-breasted Crake is a member of the rail family and a common resident throughout most of Southeast Asia, but like most rails it is not easy to spot. This attractive bird rarely flies, preferring to walk about inside patches of dense weeds and reeds growing along freshwater bodies such as rivers, lakes or swamps. In rural areas it will come onto drier terrain and overgrown fields. The early morning is the best time to see the Ruddy-breasted Crake, when it emerges briefly from its cover and runs about at the edge of the vegetation.

Family:
Rallidae

Distribution:
South and East Asia; Indonesia, Malaysia, Thailand, the Philippines and Singapore.

Like many members of its family, this crake is partly nocturnal, feeding at dusk among the reeds where it eats seeds and shoots as well as aquatic insects and their larvae. Its nest is built on the ground deep inside the reed beds and both parents tend the young. Outside of the breeding season it disperses and sometimes undertakes local migrations, probably moving by night.

Terns

Laridae is a large family that includes terns as well as gulls. However gulls are absent from tropical Southeast Asia—only a few species occur as rare migrants and none are resident. Terns are well represented and at various locations are quite plentiful. There are two types of terns—the powerful members of the *Sterna* genus (also known as sea terns) and the smaller, delicate *Chlidonias* terns (also known as marsh terns).

Among the sea terns, the most common Southeast Asian resident is the Black-naped Tern, *Sterna sumatrana*, (bottom left). This tern breeds on remote offshore islets and was formerly present in dense colonies. Lately these terns have become much reduced in numbers as local fishermen are more mobile now and visit even remote offshore reefs and rocks to collect seabird eggs. Like many other terns the Black-naped Tern does not build a nest, but lays its eggs in a depression on a rocky surface, making the eggs easy to find. Outside the breeding season these birds often live near the coastline in loose flocks and can be seen from the beach or during ferry crossings in the region.

The Little Tern, *Sterna albifrons*, (above right) is the smallest of the terns and is commonly found around remote sandy beaches, mud flats and estuaries in Southeast Asia. It is also sometimes seen inland along large river banks. All sea terns feed by diving directly into the water for small fish. The marsh terns, however, lift their food from the surface without getting wet. A migrant from temperate Asia, the White-winged Tern, *Chlidonias leucopterus*, (top left) is the most common representative of the marsh terns in Southeast Asia. It can be seen along the coasts and around inland freshwater ponds and reservoirs for most months of the year, often in large flocks numbering hundreds of birds.

Family:
Laridae

Distribution:
Asia, Australia; Indonesia, Malaysia, Thailand, Singapore and the Philippines.

Malaysian Plover

Charadrius peronii

Although a great variety of shorebirds are found along the coastlines of Southeast Asia, the Malaysian Plover is unique in being the only shorebird that is restricted to the region. Its worldwide range covers Southeast Asia, the Philippines and parts of Indochina. While other shorebirds are migratory and form dense flocks outside the breeding season, this small, enigmatic bird stays in its territory all year, usually in pairs. The Malaysian Plover is only found at white, sandy beaches, mainly in sheltered coves with washed-up coral debris—never on mudflats or around mangroves. Although fairly tame, it avoids busy places with tourists and sunseekers and can only be found along remote seashores and on offshore islands. In general it is a little-known and rarely-photographed species.

Like other members of its family, this plover runs across the flat surface of the sand, moving its legs so rapidly that it appears to "roll" along, stopping abruptly to pick up tiny insects. The nest is a depression in the sand or fine gravel just above the tide line.

Unlike many other shorebirds this species is sexually dimorphic—the adult male (above) can be easily recognized by its black markings on head and neck. These are lacking on the adult female (below).

Family:
Charadriidae

Distribution:
Southeast Asia; Indonesia, Malaysia, Thailand, the Philippines and Singapore.

Sandpipers

Family:
Scolopacidae

Distribution:
Asia;
Indonesia,
Malaysia,
Thailand, the
Philippines and
Singapore.

The sandpipers are a large family of birds, most of which breed in temperate climate zones and migrate south during the northern winter. Some nest in the high arctic areas of Siberia and move as far south as New Zealand to spend the winter. Although more than thirty species are seen regularly in Southeast Asia, sometimes in huge numbers, no members of this family breed in the region, apart from three rare woodcock genus endemic to montane parts of Indonesia and New Guinea.

All sandpipers are shorebirds, with long legs suitable for walking in water, and long bills for picking out small invertebrate prey from the mud. Sandpipers congregate during migration along shallow coastlines, especially on mudflats and around mangroves. They feed on the exposed mudflat at low tide (even if low tide occurs during the midday heat or at night), then spend the high tide period on higher ground, roosting on gravel banks or in dry fields.

Thousands of sandpipers will congregate along large expanses of mudflats. Often species are represented in separate flocks but are sometimes mixed together. Most sandpipers look very similar in winter plumage, with grey upperparts and paler underparts. Since they are often viewed at great distances, correct identification can be a challenge. On the opposite page, compare the Common Greenshank, *Tringa nebularia*, (top left) to the Marsh Sandpiper, *Tringa stagnatilis*, (top right). The Terek Sandpiper, *Xenus cinereus*, has a slightly upturned bill, while the Common Redshank, *Tringa totanus*, next to it (bottom left) has red legs. Most sandpipers prefer mudflats; however the Ruddy Turnstone, *Arenaria interpres*, (bottom right) is also found on sandy beaches, gravel banks and along concrete embankments. The Common Sandpiper, *Actitis hypoleucos*, (center) and the Wood Sandpiper, *Tringa glareola*, (above) are often found around freshwater wetlands, rivers and rice fields far from the coast. Both species are very numerous during peak migration periods.

Pigeons

Family:
Columbidae

Distributions:
Pink-necked Pigeon:
Indonesia, Singapore,
Malaysia, Thailand,
and the Philippines.

Pigeons and doves make up the large Columbidae family which includes several genera. In general the term "pigeon" is used for the larger, tree-dwelling members of the family, while doves are known as the smaller birds, most often seen feeding on the ground.

Green pigeons in the genus *Treron* are plentiful in Southeast Asia. The females are uniformly green while

the males have diagnostic patterns of brown, blue and orange in their plumage. A common garden and coastal woodland bird in the region is the Pink-necked Pigeon, *Treron vernans*, shown on the left (a male), while the Thick-billed Pigeon, *T. curvirostra*, (opposite, top left, a male) and the Little Green Pigeon, *T. olax*, (opposite, bottom right, a male) are lowland forest birds.

Imperial Pigeons in the

Little Green Pigeon,
Jambu Fruit-Dove:
Malaysia, Java, south
Thailand, Singapore,
Borneo and Sumatra.

Thick-billed Pigeon,
Green Imperial
Pigeon: South Asia;
Indonesia, Malaysia,
Thailand and the
Philippines.

Green-winged
Pigeon: South Asia;
Indonesia, Malaysia,
Thailand, Australia,
the Philippines and
Singapore.

genus *Ducula* are the largest of them all. The Green Imperial Pigeon, *Ducula aenea*, (opposite, center left) is the size of a crow. This handsome bird is common in lowland rainforest and is often seen flying rapidly above the canopies, traveling to and from fruiting fig trees. Pigeons live almost entirely on fruits. The fruit-doves *Ptilinopus* are named after this characteristic. A male Jambu Fruit-Dove, *Ptilinopus jambu*, can be seen in the photo opposite (bottom left).

The Green-winged Pigeon, *Chalcophaps indica* (also called the Emerald Dove) is considered a dove since it lives in dense forest and scrub and forages on the ground in true dove fashion. Difficult to view and photograph, it is most commonly seen flying with amazing speed, low among the forest vegetation (opposite, top right).

Spotted Dove

Streptopelia chinensis

Family:
Columbidae

Distribution:
South Asia;
Indonesia,
Malaysia,
Thailand, the
Philippines and
Singapore.

Named after the pretty markings on its neck, the Spotted Dove is found in open woodlands and cultivated areas all over Southeast Asia and is common even in villages and towns. A feral population has established itself in Australia. The Spotted Dove spends much of its time on the ground, walking about searching for grains and cereals, cooing softly. It feeds on lawns and along the roadsides where there are tall, seed-producing grasses. In cultivated areas this dove is often found near rice fields and grain storage facilities. When disturbed, it flies off a short distance and lands quickly again on the ground or a low branch. This species is a popular cage bird. Calling birds in cages are used as decoys for trapping wild Spotted Doves.

The nests of pigeons and doves are flimsy constructions—just a few sticks brought together, located low in a tree. Considering the rickety assembly it is a wonder that these birds are so widespread and numerous.

Zebra Dove

Geopelia striata

Similar in coloration and habits to the Spotted Dove, the Zebra Dove (also known as the Peaceful Dove) is a much smaller, more slender bird. Found all over Southeast Asia, it lives in open woodland, and around cultivated areas. Unlike the Spotted-necked Dove it does not live in towns and cities, but prefers the rural countryside with tree cover.

Originally the Zebra Dove had a small distribution covering the Malay Peninsula, the Philippines, Sumatra, Java and Bali, but now it has been introduced into most of Thailand, Borneo, Sulawesi and other parts of the region. Paradoxically, the decline in the number of this species within its original localities is due to the same trapping for the caged bird trade that has caused its range to expand.

The Zebra Dove has a soft call consisting of 6–8 rolling whistles, a song not all that exciting to most people. However the bird is highly prized among songbird fanciers; individuals doing well in bird-singing contests fetch throusands of dollars in the trade.

Family:
Columbidae

Distribution:
Southeast Asia and Australia; Indonesia, Malaysia, Thailand, the Philippines and Singapore.

Blue-crowned Hanging Parrot

Loriculus galgulus

Family:
Psittacidae

Distribution:
South Thailand,
Malaysia,
Singapore,
Borneo and
Sumatra.

Several tiny, green *Loriculus* parrots occur in Southeast Asia. The Philippines has one endemic species and Indonesia has several in the eastern part of the archipelago. A central Southeast Asian species, the Blue-crowned Hanging Parrot is found in Malaysia, Singapore, southern Thailand and parts of Indonesia.

This parrot is a lively and attractive bird to watch, although it is sometimes difficult to see well, as it often perches high in the canopy of lowland rainforest trees feeding on small fruits, new shoots and flowers. As its name indicates, the parrot will often clamber through the branches head down and reportedly even sleeps this way, hanging by its legs like a bat. More often this diminutive 12-cm bird is seen flying past in small flocks just above the forest canopy. The flight pattern is rapid and direct with whirling wingbeats. This species keeps in contact with a characteristic, high-pitched, ringing call. A male is shown below—the female lacks the red spot on the chest.

Long-tailed Parakeet

Psittacula longicauda

Parrots are not a principal family of birds in Southeast Asia—only in eastern Indonesia and Australia do they dominate the avifauna. Although only a few species occur in this region, the slender, long-tailed parakeets of the genus *Psittacula* are fairly easy birds to find. These parakeets all live in lowland forest edges, coastal woodlands and around plantations and parks. Locally they can be quite common and make their presence known with a loud, screeching call that carries a long way. All parakeets are conspicuous birds, often seen in small compact flocks, flying rapidly above the woods, eventually landing in a fruiting or flowering tree to feed. At dusk the area's population will congregate on large communal roosts.

Parakeets nest in a cavities in trees—often in an abandoned woodpecker's nest or a rotten stump, as they are not able to make their own holes. A male Long-tailed Parakeet is shown below. The female parakeet has much paler black and red face colors.

Family:
Psittacidae

Distribution:
Malaysia,
Singapore,
Borneo and
Sumatra.

Chestnut-bellied Malkoha

Phaenicophaeus sumatranus

Family:
Cuculidae

Distribution:
South Thailand,
Malaysia,
Singapore,
Borneo and
Sumatra.

Malkohas are members of the cuckoo family, but unlike the smaller "true cuckoos" they are not nesting parasitic. The genus is characteristic of a number of species found in lowland rainforest habitat in Southeast Asia. You cannot spend a morning birding in the rainforest without seeing one or more malkohas. The Chestnut-bellied Malkoha is common locally in primary and secondary lowland rainforest and sometimes in mangroves.

The malkoha is a large, long-tailed bird with secretive habits that moves about inside the top forest canopy and the middle storeys of large trees, jumping along the branches like a mammal. This bird flies only briefly when it has to move to another tree, gliding much of the way.

Usually found in pairs, the Malkohas constantly search for large insects, especially cicadas. Their nest is built with small sticks high in a forest tree, but little more is known about their nesting habits—as is the case with so many Southeast Asian forest birds.

Indian Cuckoo

Cuculus micropterus

Cuckoos are well represented in Southeast Asia. This species is distributed as resident and/or migrant throughout the whole region. The Indian Cuckoo is very difficult to distinguish in the field from the Oriental Cuckoo, *C. saturatus*, which is also a migrant and a resident at higher altitudes in Southeast Asia. The *Cuculus* cuckoos are all nesting parasitic and especially favor members of the warbler family Sylviidae for hosts.

Family:
Cuculidae

Distribution:
South Asia; Indonesia, Malaysia, Thailand, the Philippines and Singapore.

The Indian Cuckoo, a shy forest bird, often moves high among the treetops in lowland rainforest, and so is difficult to see clearly. As it flies between the trees on narrow pointed wing's, it looks like a small bird of prey. Its call can often be heard during the breeding season—a loud, clear four-part whistle with the fourth stroke lower, something like Beethoven's 5th Symphony! This call is repeated over and over and is usually the best way of locating this species. During migration the Indian Cuckoo can be found in a wider variety of woodlands and forested areas.

Lesser Coucal

Centropus bengalensis

Family:
Cuculidae

Distribution:
South Asia;
Indonesia,
Malaysia,
Thailand, the
Philippines and
Singapore.

Coucals are large members of the cuckoo family. This species, the most common, is found all over Southeast Asia in open country, scattered woodlands and marshy areas with plenty of tall grasses. Care should be taken to distinguish it from the Greater Coucal, *C. sinensis*, which is larger, lacks pale spots in the brown and black plumage and is found in more forested terrain.

The Lesser Coucal is a peculiar bird seen best early in the morning, when it often emerges to rest in the open, perched on a clump of grasses or a low branch. The rest of the day is spent deep inside the long grass where it walks on the ground like a mammal, searching for insects, frogs, lizards and other small prey. It makes short, low flights with shallow wing beats and glides briefly before it dives back into cover. The Lesser Coucal, not nesting parasitic, builds its own nest near the ground. The call is a distinct series of deep, whooping notes, often heard resounding over the open grasslands and marshes.

Plaintive Cuckoo

Cacomantis merulinus

This small cuckoo, widely distributed throughout Southeast Asia, is quite common locally, yet not often seen. The Plaintive Cuckoo is somewhat similar to its close relative, the Rusty-breasted Cuckoo (or Brush Cuckoo), *C. sepulcralis*, which is more of a forest bird, with a rust-colored breast and throat and a distinct, yellow eye-ring.

The Plaintive Cuckoo prefers a habitat of open woodlands, disturbed forest edges and cultivated areas, but can also be found in mature village gardens. It feeds on insects and other invertebrate prey and sometimes on small fruits.

This cuckoo is somewhat anonymous in appearance and behavior, but like other small cuckoos it has a very distinctive call—a mournful, monotonous whistle. It lays its eggs in the nests of other smaller birds, parasiting on their parental efforts. The cuckoo is often mobbed on its perch during the day by these smaller birds. When it finally has had enough, the cuckoo will fly off abruptly, quickly disappearing among the branches of another tree.

Family:
Cuculidae

Distribution:
South Asia;
Indonesia,
Malaysia,
Thailand, the
Philippines
and Singapore.

Collared Scops-Owl

Otus lempiji

Family:
Strigidae

Distribution:
South Asia;
Indonesia,
Malaysia,
Thailand, the
Philippines and
Singapore.

Owls are plentiful in Southeast Asia. A number of large and small species are well represented, but by their very nature are difficult to observe. All species are nocturnal, most are dark brown and live in remote parts of dense forests. Night outings can be very rewarding—apart from the owls you may see nightjars, nocturnal mammals, insects, reptiles, amphibians, self-illuminating fungi and fireflies which occur around the mangrove forests.

The Collared Scops-Owl is one of the owls that you are most likely to see in the region. This is a forest bird but can also be found along the forest edges, around plantations and even in parks and villages. Its call is a single, soft hoot repeated every 10 seconds or so, especially right after dark in the breeding season. The owl hunts from a low perch taking insects, small reptiles and mice from the ground. The nest is built in a cavity in a tree. The adult individual shown below is returning to the nest with a cricket for its young.

Large-tailed Nightjar

Caprimulgus macrurus

Nightjars are nocturnal birds that feed on insects. Several species occur in Southeast Asia. Some are forest birds; others, like this species, come out to forest edges and into more open country. The Large-tailed Nightjar is widespread and common throughout Southeast Asia. It spends the day resting on the ground or sitting on a branch low in a tree. Emerging at dusk it flies about, a short distance above the ground, swerving from side to side catching flying insects with its small beak that opens up into an enormous gape. Often found close to human habitation, it is seen chasing moths attracted to village streetlights, especially after heavy afternoon rains. Its peculiar unbird-like call, a monotonous, metallic "chunck-chunck-chunck", is repeated continuously for half an hour or more during the night to mark its territory. The Large-tailed Nightjar lays its eggs on the ground in a small depression, usually under a tree or a bush, with no attempt at nest-building. The nightjar shown below is sitting on her eggs.

Family:
Caprimulgidae

Distribution:
South Asia;
Indonesia,
Malaysia,
Thailand, the
Philippines and
Singapore.

Kingfishers

Family;
Alcedinidae

Distribution:
South Asia,
Indonesia,
Malaysia,
Thailand, the
Philippines and
Singapore.

This family of brightly-colored birds occurs throughout Southeast Asia. Metallic blue and orange colors dominate their plumage and all have strong beaks used for capturing prey. Some dive into water for fish but most live on insects and small vertebrates. Kingfishers perch upright on poles or branches and all have loud and somewhat harsh calls often heard when they take wing. Their flight is rapid and direct, usually low and close to the ground or over the surface of the water. The nest is always in a cavity—often a burrow in a small hill, a river bank or sometimes a hole in a tree.

Kingfishers belonging to the genus *Halcyon* are large, chunky birds. Two species are particularly prominent residents of Southeast Asia. The Collared Kingfisher, *Halcyon chloris*, (opposite, top right) is mostly found along the coast in mangrove forests, coastal woodlands and sometimes around parks and gardens. The White-throated Kingfisher, *H. smyrnensis*, (above) is often seen along rivers, in marshy areas and sometimes in open woodlands and around cultivated areas. Both are highly territorial—often a pair will visit the same perch throughout the year.

Alcedo kingfishers are small birds that dive into the water for fish. The Common Kingfisher, *Alcedo atthis*, (left) is a familiar winter visitor to wetlands all over Southeast Asia, staying to breed in some areas. The Blue-eared Kingfisher, *A. meninting*, (opposite, top left) is strictly a forest bird found near remote rivers and streams upcountry.

Also a forest species, the Oriental Dwarf Kingfisher, *Ceyx erithacus*, (opposite, bottom left) is similar to the *Alcedo* kingfishers, although placed in a genus of its own. The Banded Kingfisher, *Lacedo pulchella*, (opposite, bottom right) is different again, strictly a lowland rainforest bird, often found sitting motionless in the middle storey of the rainforest, far from wet areas.

34

Diard's Trogon

Harpactes diardii

Family:
Trogonidae

Distribution:
South Thailand,
Malaysia, Borneo
and Sumatra.

Trogons are true denizens of lowland and lower montane rainforest—beautiful and unobtrusive birds found in both primary forest and mature secondary growth. These members of the Trogonidae family are medium-sized birds, with long tails and short, rounded wings used to maneuver silently through the rainforest. All species in Southeast Asia are colorful, usually with red or orange bellies, but are surprisingly difficult to spot. Learning to recognize the various species' territorial calls is an important aid in surveying for this family and many other forest birds.

The Diard's Trogon is a common bird locally but not numerous anywhere. Trogons usually sit motionless on a perch below the canopy in dense forest, waiting for insects to fly by or crawl past on the foliage, and will often suddenly vanish from sight. Trogons are presumed to nest in cavities in trees but few nesting records of this species exist. A male is shown (bottom left) and a female (bottom right).

Bee-eaters

Bee-eaters are some of the most attractive and colorful birds in Southeast Asia. All species have greenish plumage and long, slender bodies, long tails and pointed wings. From a perch out in the open—on a dead branch, pole or a wire—they fly out to capture bees and other insects in their long beaks with a snap that you can often actually hear.

Two species are common. The Blue-tailed Bee-eater, *Merops philippinus*, (bottom left) breeds in the northern part of Southeast Asia, but is an abundant winter visitor throughout the region. During peak migration periods it seems to be everywhere in open country, forming gigantic flocks at evening roosting sites near the coast. The Blue-throated Bee-eater, *M. viridis*, (bottom right) is resident all over Southeast Asia except in east Java and Bali. A breeding colony of this species is one of the great birdwatching spectacles, as dozens of birds busily dig borrows into a beach or river bank while others glide elegantly to and fro, filling the air with their soft, ringing calls.

Family:
Meropidae

Distribution:

Blue-tailed Bee-eater: South Asia; Indonesia, Malaysia, Thailand, the Philippines and Singapore.

Blue-throated Bee-eater: Southeast Asia; Indonesia, Malaysia, Thailand, the Philippines and Singapore.

Hornbills

Since hornbill images are often used in tourist promotion, this group is probably the most widely recognized family of birds in the region. All Southeast Asian hornbills are massive birds with large bills adorned by a characteristic casque. Hornbills are mainly fruit eaters and can be seen flying high across the rainforest, traveling towards fruiting fig trees. Several species are often seen feeding together and some congregate at evening roosts outside the breeding season. Recent research has shown that the hornbill diet also includes invertebrates, reptiles and even some birds and mammals. In general these birds are quite easy to find in primary lowland and lower montane rainforest and sometimes in mature secondary growth if primary forest is nearby. They are not easy to see clearly, as they tend to perch high in the forest at canopy level.

Twenty species occur in Southeast Asia. Some are widespread throughout the region and others are endemic to islands in the Philippines and Indonesia. Several are in danger of global extinction. All depend on large hardwood trees in mature forests for nesting. The male seals the female inside a cavity, high in a live tree, using regurgitated fruits, feces and mud for plaster. Here she stays until the young is hatched and sometimes until the young is ready to fly. During that period the male visits her six to seven times a day to bring food.

Shown above left is the Wreathed Hornbill, *Rhyticeros undulatus*, a male. The photographs on the opposite page show the Great Hornbill, *Buceros bicornis*, a male (top left); the Oriental Pied Hornbill, *Anthracoceros albirostris*, a male (bottom right); the Rhinoceros Hornbill, *Buceros rhinoceros*, male (top right); and the Brown Hornbill, *Ptilolaemus tickelli*, two males near a nest (bottom left).

In recent years Asian hornbills have been the focus of several expeditions and research projects. This exciting work brings new findings to light every year.

Family:
Bucerotidae

Distributions:

Wreathed Hornbill, Great Hornbill and **Oriental Pied Hornbill:** South Asia; Indonesia, Malaysia and Thailand.

Rhinoceros Hornbill: Malaysia, Sumatra, south Thailand and Borneo.

Brown Hornbill: Parts of South Asia; Thailand and Indochina.

Photo above by Eric Woods.

Barbets

Family:
Megalaimidae

Distributions:

Red-throated Barbet:
South Thailand,
Malaysia, Borneo and
Sumatra.

Blue-eared Barbet:
Southeast Asia;
Indonesia, Malaysia
and Thailand.

Most species in this small family of tropical birds live in Southeast Asia. A morning walk in the rainforest is not complete without seeing one or more barbets, usually hopping about high inside the trees at canopy level. Barbets are related to woodpeckers. They have strong bills and usually sit across the branches or vertically on tree trunks as woodpeckers do. Their nests are built high in trees, in cavities that they excavate themselves.

Barbets live almost entirely on fruits—a fig tree in fruit is sure to attract several species. Most barbets have green plumage with colorful patterns on head and throat which are often diagnostic. You will always hear the barbets—all have repeated hooting and clucking calls that would do well as jungle backdrop sounds in a Hollywood movie.

The Red-throated Barbet, *Megalaima mystacophanos,* (below left) is a fairly large species found only in primary and secondary lowland rainforest. The Blue-eared Barbet, *M. australis,* (below right) occurs in montane habitats up to 1500 meters elevation and has a wider distribution including parts of India and southern China. Both existed previously in Singapore, but are now locally extinct.

Crimson-winged Woodpecker

Picus puniceus

Woodpeckers belong to a large family of birds distributed on all continents except Australia. The birds in this family are arboreal and depend on the rainforest or the forest edges and mangrove habitats for food and shelter, although a few species will move into nearby gardens and parks. Woodpeckers live mainly on ants and insect larvae that they pry out of the bark with their strong bills. Their bills are also used to obtain sap from trees and to excavate their nesting holes, often in dead parts of trees where the wood is softer. You will hear them tapping in rapid succession as part of their characteristic territorial display.

The Crimson-winged Woodpecker, common locally in lowland primary and secondary rainforests, is a typical, medium-sized member of the Picidae family. However this bird and nine other members of its family are now considered extinct in Singapore.

It is often seen flying from tree to tree in the rainforest, landing low and exploring the trunk and branches for food as it moves gradually up to the canopy where it glides quickly to another tree. The Crimson-winged Woodpecker (female) shown on the left sits along the branch and supports herself with her strong tail.

Family:
Picidae

Distribution:
South Thailand, Malaysia, Borneo, Sumatra and Java.

Long-tailed Broadbill

Psarisomus dalhousiae

Family:
Eurylaimidae

Distribution:
South Asia;
Indonesia,
Malaysia and
Thailand.

The broadbills are a family of peculiar arboreal birds—a small group of specialist species—most of which live in Southeast Asia. These birds adapt poorly to disturbance of their rainforest homes. Previously five species were found in Singapore but are now all locally extinct. These species can still be found in neighboring countries, but most are rare. Seeing a broadbill during a morning walk is always a thrill.

Most broadbills are brightly colored, but are not easy to observe since they are small- to medium-sized birds that move in the top canopies or just below in the mid-storey section of the forest. Here they live mainly on insects and sometimes fruits. The Long-tailed Broadbill is best located by its loud, 5- to 7-note, whistling call. A submontane species, found in primary rainforest between 700 and 2000 meters elevation, this broadbill does not seem to be numerous anywhere, except in Thailand where it can be locally common.

Blue-winged Pitta

Pitta moluccensis

Found throughout Southeast Asia, this pitta is one of the most widespread members of its family. Resident in parts of Indochina, Thailand, Malaysia and Indonesia this bird is a migrant and winter visitor in the rest of the region.

Although usually a forest ground bird, during migration the Blue-winged Pitta can also be found in disturbed habitats, including scrub and parklands, often near the coast. Like all pittas it is shy, moving along the ground with long, hopping strides, stopping to peck for insects and other invertebrates in the soil. It perches on low branches and if disturbed flies off moving its small wings rapidly in a direct whirling flight, keeping low to the ground.

The nest of the Blue-winged Pitta is a ball of vegetation built near or on the ground. Like other members of the family it has a soft characteristic whistle which is often the only evidence of its presence. Pittas are usually difficult to locate and observe and getting good photographs of one is regarded as the ultimate prize of a birding trip.

Family:
Pittidae

Distribution:
South Asia; Indonesia, Malaysia, Thailand, the Philippines and Singapore.

Pacific Swallow

Hirundo tahitica

Family:
Hirundinidae

Distribution:
South Asia and
the Pacific
Islands;
Indonesia,
Malaysia,
Thailand, the
Philippines and
Singapore.

Swallows and swifts (family Apodidae) are aerial feeders, however swifts usually fly high in the air, moving fast and gliding on extended wings, while swallows fly low near the ground. Swifts spend practically all their time in the air and only land when they fly into a cave to nest. Swallows frequently perch on branches, wires and poles.

This species is the most widespread and numerous of the swallows here and is a common resident throughout Southeast Asia. Its range extends from India all the way to the islands of the Pacific. In size and behavior this swallow is similar to the Barn Swallow, *Hirundo rustica*, which is found almost worldwide and occurs as a migrant and winter visitor in this region. However the Pacific Swallow can be recognized in all plumages by its pale breast and shorter, less forked tail. Outside the breeding season the two species often move about together, sometimes gathering in massive flocks numbering thousands of birds, congregating late in the day at certain communal roosting sites on buildings or in reed beds.

The Pacific Swallow is gregarious, always seen in small flocks, often near the coast or other wetlands, flying about continuously. It is a fast and elegant flier, constantly twisting and turning in the air, to catch minute insect prey. When resting, the swallow sits on a low perch singing quietly. The Pacific Swallow has a pleasant and varied little twittering call.

Swallows build their nests on vertical surfaces from small pellets of mud collected from nearby puddles and river banks. Since swifts do not land on the ground, they use saliva and feathers for their nests. In fact some species' nests are considered edible and a delicacy. Swallow nests are found under rocky cliff overhangs, but just as often under the eaves of buildings or concrete bridges and dams.

Leafbirds

Leafbirds are a small family of forest birds that you can only see in tropical Asia. These are active and conspicuous birds, and easy to observe, even though they usually move high in the trees. They are mostly green, although some males have yellow, blue and black patterns on their heads and breasts. Leafbirds live on insects and fruits growing in the canopies of large trees. Sometimes they come down briefly to the lower stories, often hopping about at the edge of the foliage while calling with loud, clear whistles.

Several species do well in the primary and secondary rainforest of Southeast Asia, both in the lowlands and at lower montane elevations. Three species are still holding on in Singapore's forest reserves.

The Greater Green Leafbird, *Chloropsis sonnerati*, is recognized by its size and heavy bill—a female is shown below. The Blue-winged Leafbird, *C. cochinchinensis*, is the most common of the leafbirds in the lowlands. The leafbird shown above is a male of this species.

Family:
Chloropseidae

Distributions:
Greater Green Leafbird: South Thailand, Malaysia, Singapore, Borneo, Sumatra, Java and Bali.

Blue-winged Leafbird: South Asia; Indonesia, Malaysia, Thailand and Singapore.

Yellow-vented Bulbul

Pycnonotus goiavier

Family:
Pycnonotidae

Distribution:
Southeast Asia;
Indonesia,
Malaysia,
Thailand, the
Philippines and
Singapore.

This attractive bird, perhaps one of the most characteristic and common of all, is distributed throughout the region. Although originally a mangrove and coastal woodlands resident, the Yellow-vented Bulbul has now adapted totally to the human presence. Common in all kinds of open woodlands, gardens and parks, it has even been known to nest in potted plants on urban, high-rise balconies. Like most successful birds it is omnivorous, feeding on fruits, seeds, nectar and insects which it gleans from the foliage, catches in the air or finds on the ground. Its song is a rich, bobbling chatter that can be heard in any Southeast Asian village, especially at daybreak when there are few other sounds around. Usually it moves about singly or in pairs, although small flocks tend to form in fruiting or flowering trees. The nest is a cup located low in a tree or dense bush. Although this is the only bulbul found in gardens, the family is otherwise well represented in the region and many more species live in lowland rainforest.

Greater Racket-tailed Drongo

Dicrurus paradiseus

At first glance the drongos may not look all that attractive, as they are uniformly black or dark grey birds. However they are a pleasure to observe, being active and elegant in the air, often flying out and calling from an open perch. This species has two exotic-looking, elongated tail-streamers, although sometimes one or both of these "rackets" might be missing on some individuals. The Greater Racket-tailed Drongo lives in trees in primary forest, mature logged forest and nearby cultivated areas. From its perch it hawks for air-borne insects, like a flycatcher. The drongo has a bizarre call and can produce a wide variety of sounds including melodious whistles and metallic noises mixed with imitations of other birds' calls. The nest is a small cup built in thin branches high in a tree and the long tail feathers hanging over the edge make the sitting bird easy to spot. This drongo is locally common in forested parts of the region including Indochina and Singapore. The Lesser Racket-tailed Drongo, *D. remifer*, is restricted to montane forest.

Family:
Dicruridae

Distribution:
South Asia;
Indonesia,
Malaysia,
Thailand and
Singapore.

Crows

Family:
Corvidae

Distributions:

Large-billed Crow: South Asia; Indonesia, Malaysia, Thailand, the Philippines and Singapore.

House Crow: South Asia; Malaysia, Thailand and Singapore.

Crows are intelligent, adaptable birds and many species do very well in disturbed environments. The Large-billed Crow, *Corvus macrorhynchos*, (above) is the most widespread and common of the crows found throughout the region—in woodlands, open country and near villages and towns. It does not enter cities however and remains fairly shy. Outside the breeding season the crows of the area congregate at conspicuous roosting sites, often in coastal mangrove trees. The crow is unique in its ability to transcend altitude boundaries and is found from coastal lowlands all the way up to the highest mountain plateaus. It is omnivorous, feeding on all edible matter—both vegetable and animal, including carrion and garbage.

Locally even more successful, the House Crow, *Corvus splendens*, (below) seems to displace the Large-billed Crow wherever it settles. Introduced by sailors from south India, this greyish species has established itself largely in Singapore and parts of Malaysia.

Black-naped Oriole

Oriolus chinensis

This handsome bird is widely distributed throughout tropical Southeast Asia and can be fairly common locally. The male (shown below) has especially stunning plumage. The Black-naped Oriole is arboreal, but not a forest bird, preferring open coastal woodlands and mangrove fringes for habitat. This bird has adapted well to disturbed environments and is found in village parks and gardens, even some city centers as long as big ornamental trees are available. Singapore is a good place to see this species.

The Black-naped Oriole hops about in search of fruits and large insects, but usually stays inside the canopy of the tree being a somewhat timid creature. Most resident tropical birds are active in the morning, especially from eight to ten o'clock. However this species becomes very conspicuous again in the evenings before dusk when it can often be seen flying restlessly among the scattered trees, a few birds displaying together and calling loudly with a characteristic, clear, fluty whistle.

Family:
Oriolidae

Distribution:
South Asia; Indonesia, Malaysia, Thailand, the Philippines and Singapore.

Babblers

Family:
Timaliidae

Distributions:

Striped Tit-Babbler:
South Asia;
Indonesia,
Malaysia, Thailand,
the Philippines and
Singapore.

**Sooty-capped
Babbler**: South
Thailand, Malaysia,
Borneo and
Sumatra.

**Scaly-crowned
Babbler**: South
Thailand, Malaysia,
Borneo, Sumatra
and Java.

Silver-eared Mesia:
South Asia;
Thailand,
Peninsular
Malaysia,
and Sumatra.

This is by far the largest of all bird families in Southeast Asia. When Ben King wrote his *Field Guide to the Birds of South-East Asia* in 1975, he listed 139 species of babblers for the region—no other family comes close to 100 members here. All babblers are insect eaters. Most are fairly small, dull-colored birds which live in forested areas, usually close to the forest floor. They never fly long distances and none of the species migrate.

Being restless and skulking birds, babblers are typically difficult to view clearly and in the rainforest the many different (but similar) species will provide some real brainteasers—even for the experienced birdwatcher. Babblers have a variety of calls, from clear whistles to harsh, scratching noises. Often calls are the best instrument when attempting to locate and identify these elusive creatures.

Only a few of the many babbler species are widespread and common throughout the region. The Striped Tit-Babbler, *Macronous gularis*, (top) is one of these, common in all countries in diverse habitats—from primary lowland forest to secondary forest edges and scrub. Most other babblers are rainforest specialists. Some, like the members of the *Malacopteron* genus— the Sooty-capped Babbler, *Malacopteron affine*, (bottom left) and the Scaly-crowned Babbler, *M. cinereum*, (bottom right)—are only found in a narrow niche of lowland forest undergrowth, from the floor up to the lower mid-story. Other species are endemic to a certain area, for instance the Sunda Laughing Thrush, *Garrulax palliatus*, is found only on Sumatra and Borneo. Many babblers are restricted to montane habitats, especially to the lower montane zone from 600–2000 meters a.s.l. which is rich in invertebrates and insectivorous wildlife. The Silver-eared Mesia, *Leiothrix argentauris*, (above) is one such species, found only in elevated rainforest where it can be quite common locally.

Magpie Robin

Copsychus saularis

Family:
Turdidae

Distribution:
South Asia;
Indonesia,
Malaysia,
Thailand,
the Philippines
and Singapore.

Widespread throughout Southeast Asia, the Magpie Robin occurs in all kinds of habitats from forest edges to coastal woodlands and mangroves. This bird has adapted well to the park and garden environment and can often be heard singing from fence poles and roof tops in villages. In a region where birds are not usually known for being tame, this trusting bird stands out with its friendly behavior. It perches low and often feeds on the ground like other members of the thrush family, hopping about looking for small invertebrates in the grass, lifting and lowering its tail.

Its song is a melodious whistle, in fact so melodious that this bird is caught and kept as a caged pet. Although the species is still common in most countries, this practice has led to declining populations in many areas. In Singapore the Magpie Robin was virtually extinct for some years but now seems to be recovering slowly. The Magpie Robins shown below are both males. The subspecies found on eastern Java and Bali have all-black underparts.

White-rumped Shama

Copsychus malabaricus

Plenty of sounds can be heard in the rainforest—made by birds, insects and amphibians. But the chorus consists mainly of chipping and rattling noises—there are few real songsters among these animals. The Shama's song, however, is a delight to hear—luckily it is not an uncommon bird. The Shama is found all through Southeast Asia, in lowland rainforest and along the forest edges. Its song is powerful and melodious and often the bird sings late in the morning or even in the middle of the day, when there are few other sounds around.

However the Shama is not that easy to see. It sings from a low perch, but if you try to approach, the bird will stop singing and all you see when it flies away is an orange flash, a long tail and a bright, white rump patch disappearing among the trees. Then the song starts again from a new, hidden location further inside the forest. A popular cage bird, this species is a valuable commodity in the wildlife trade business. Males of this species are shown below.

Family:
Turdidae

Distribution:
South Asia;
Indonesia,
Malaysia,
Thailand and
Singapore.

Thrushes

Family:
Turdidae

Distributions:

Orange-headed Thrush: South Asia; Indonesia, Malaysia, Thailand and Singapore.

Chestnut-capped Thrush: Southeast Asia; Indonesia, Malaysia, the Philippines and Thailand.

The Turdidae or thrush family is large and diverse. These forest birds perch low in the trees and spend much of their time on the ground, where they hop around searching among the dead leaves for insects and fallen fruits. They tend to be shy and are not often seen.

Members of the genus *Zoothera* are among the so-called true thrushes. The Orange-headed Thrush, *Z. citrina*, is widespread throughout Southeast Asia, resident mainly in the northern areas and migrant throughout the rest of the region (below right, a male). This is not an easy bird to see, as it does not seem to be common anywhere and in spite of its bright plumage it melts easily into the surrounding undergrowth. The Chestnut-capped Thrush, *Z. interpres,* (below left) is a shy bird and unlike most true thrushes is sedentary and does not migrate. Since this thrush requires a large expanse of primary lowland rainforest to survive, it has a small distribution. It is seldom observed during surveys and little is known of its status and habits.

Ashy Tailorbird

Orthotomus sepium

A coastal mangrove bird, the Ashy Tailorbird is also commonly found in nearby scrub and gardens on Java and Bali. This bird is restless, vocal and constantly moving, usually seen looking for insects low in the bushes and occasionally higher in the trees. A separate subspecies lives in a variety of wooded habitats, from the lowlands up to the mountain slopes.

The tailorbirds (genus *Orthotomus*), a group of Oriental region birds, have a unique way of building their nests. These birds sew large leaves together with fibers and spider web, then construct a tiny cup-nest under this shelter—hence their name. Members of the warbler family, tailorbirds are small insectivorous birds most often found in open woodlands and marshy areas. A handful of species occurs in Southeast Asia. The family is dominant in temperate climate zones, but less so in tropical areas. Here other families—the babblers especially—take over the warblers' niche in the rainforest.

Family:
Sylviidae

Distribution:
Southeast Asia;
Indonesia,
Malaysia,
Thailand, the
Philippines and
Singapore.

ALAN OWYONG

Asian Paradise-Flycatcher

Terpsiphone paradisi

Family:
Monarchidae

Distribution:
South Asia;
Indonesia,
Malaysia,
Thailand and
Singapore.

Most flycatchers in Southeast Asia belong to the Muscicapidae family but the Monarchs are now treated as their own family, Monarchidae. These are fairly large birds that sally forth for insects from an open perch like other flycatchers but also glean insects and their larvae from the foliage. This species is widely distributed and one of the most commonly seen of the family in the region. A resident and migrant throughout, it breeds inside primary and mature secondary rainforest. Outside the breeding season it will undertake migratory travels and then might also emerge in disturbed wooded habitats and even gardens close to forested areas.

Most Asian Paradise-Flycatchers look like the bird below. All females, immatures and molting males have this plumage, with the exception of older males that develop a long set of tail feathers. Some males occur in a white morph—the chestnut brown is replaced by white. Then the bird looks quite unusual as it flutters through the dense forest, the long tail dragging behind like a small, white ghost.

Shrikes

Seven species of shrike are found in Southeast Asia. These are medium-sized birds with large heads and strong bills that live on animal prey—mostly insects, small reptiles, rodents and other birds. In other regions shrikes often live near thorny bushes and impale their prey on the spikes for later consumption, but this rarely seems to be the case in the Southeast Asian tropics.

Most shrikes, such as the widespread and resident Long-tailed Shrike, *Lanius schach*, (below right) are open country birds. For most of the day they can be seen pouncing on prey in the grass below from their perch on low branches or poles, or more rarely, hawking for insects in the air—even during the midday heat. The Tiger Shrike, *L. Tigrinus*, (below left) prefers a more closed habitat. This shrike is most often found along forest edges and in denser wooded areas and sometimes in scrub and gardens. It is a migrant that breeds in China and other parts of East Asia. During the northern winter this quiet and inconspicuous bird converges on Southeast Asia where it becomes quite common.

Family:
Laniidae

Distributions:

Long-tailed Shrike: South Asia; Indonesia, Malaysia, Thailand, the Philippines and Singapore.

Tiger Shrike: East Asia; Indonesia, Malaysia, Thailand, the Philippines and Singapore.

Starlings and Mynas

Family:
Sturnidae

Distributions:

White-vented Myna: South Asia; Indonesia, Malaysia, Thailand and Singapore.

Common Myna: South Asia; Indonesia, Malaysia, Thailand and Singapore.

Philippine Glossy Starling: Southeast Asia; Indonesia, Malaysia, Thailand, the Philippines and Singapore.

Hill Myna: South Asia; Indonesia, Malaysia, Thailand, the Philippines and Singapore.

Starlings and mynas are among the world's most successful birds. The story of the pair of European Starlings, *Sturnus vulgaris*, released in Central Park in New York City 100 years ago, is legendary. The pair went on to colonize the entire North American continent. Similarly, in Southeast Asia, the White-vented Myna, *Acridotheres javanicus*, (a bird that occurs naturally in subtropical Asia, Thailand, Indochina and on Java and Bali) was introduced into Singapore in the 1930s. The myna quickly multiplied and is today the republic's most numerous bird by far and has since spread from Singapore into Peninsular Malaysia.

The Common Myna, *Acridotheres tristis*, is still the dominant species throughout much of the region and has established itself, from captive populations, in new places, mostly around villages and towns. These two species are shown together (top), the Common Myna is on the right. Both mynas eat anything edible and adapt well to the human presence. Since they thrive in disturbed habitats, they have benefited from the clearance of forests in the region.

The Philippine Glossy Starling, *Aplonis panayensis*, (bottom right) is another successful member of this family. Widespread and numerous throughout the region, it feeds mainly in trees—especially fruiting figs and ornamental trees and bushes. This starling is present in all habitats except forests, nesting in cavities in buildings and trees. Like all starlings it is gregarious even during the nesting season, especially at night when large flocks congregate at roosting sites.

Most mynas are open woodland birds, however the Hill Myna, *Gracula religiosa*, (bottom left) is an exception. Its penetrating whistle can be heard in lowland primary rainforest, mature secondary rainforest and along forest edges with large trees. This bird is also sometimes found in coastal coconut groves near forested areas. Although widespread throughout Southeast Asia, it has never become really numerous anywhere.

Sunbirds

Flowers are an important food source for many birds in tropical Southeast Asia. Bulbuls, parrots, orioles, white-eyes and starlings all visit flowering trees regularly. Members of the sunbird family live almost exclusively on flower nectar. These birds spend all day flying from flower to flower, inserting their long tongues and extracting nectar. In the process some species also pick out a few tiny insects and little fruits, but this is secondary nourishment. Nectar-eaters such as insects, bats and sunbirds participate in the pollination of many trees and bushes on their rounds. Sunbirds often have a small, favorite patch of ornamental flowers in a garden or park that they visit regularly each morning.

The two most widespread and common sunbirds in the region are the Olive-backed Sunbird, *Nectarinia jugularis*, and the Brown-throated Sunbird, *Anthreptes malacensis*. Originally coastal woodland and mangrove species they have adapted well to human disturbance and have multiplied. The Olive-backed Sunbird, the smallest of the two birds, measuring 11 cm, is very slender and swift. In fact it is sometimes mistaken for a hummingbird, a family of birds not represented anywhere in Asia. A male of this species is shown on this page, above left. The Brown-throated Sunbird is slightly larger (14 cm) and has a stronger bill. The male (opposite, bottom right) is more colorful than the female (opposite top right).

A garden in Southeast Asia is incomplete without regular visits from these birds. The smaller species will enter urban environments, however the larger sunbird is more common in rural areas. Less common but just as spectacular is the Crimson Sunbird, *Aethopyga siparaja*, (top left) a male. Widespread in the region, it prefers rural areas and gardens and parks near forest edges. A few species of sunbird, such as the Ruby-cheeked Sunbird, *Anthreptes singalensis*, (bottom left) are found mainly in rainforest. Like most rainforest birds, these species do not have the ability to adapt to disturbed habitats such as parks and gardens.

Family:
Nectariniidae

Distributions:

Olive-backed Sunbird:
Southeast Asia, Australia; Indonesia, Malaysia, Thailand, the Philippines and Singapore.

Brown-throated Sunbird:
Southeast Asia; Indonesia, Malaysia, Thailand, the Philippines and Singapore.

Crimson Sunbird:
South Asia; Indonesia, Malaysia, Thailand, the Philippines and Singapore.

Ruby-cheeked Sunbird:
Southeast Asia; Indonesia, Malaysia and Thailand.

Flowerpeckers

Family:
Dicaeidae

Distributions:

**Orange-bellied
Flowerpecker**:
South Asia;
Indonesia,
Malaysia,
Thailand, the
Philippines and
Singapore.

**Yellow-breasted
Flowerpecker**:
South Thailand,
Malaysia, Borneo
and Sumatra.

Flowerpeckers have short bills unsuitable for penetrating deep into the flowers—consequently they feed on seeds, little fruits and insects, as well as nectar. Several species are associated with mistletoes and act as important seed dispersal agents for this group of plants. Flowerpeckers are tiny birds—the Orange-bellied Flowerpecker, *Dicaeum trigonostigma*, (below right) is the smallest in Southeast Asia, with a length of just 8 cm.

This family consists mainly of forest birds, although many, like the Yellow-breasted Flowerpecker, *Prionochilus maculatus*, (below left) occur both in primary forest and along disturbed forest edges. Unlike most other forest birds which specialize in a certain level of the tall forest, flowerpeckers move across a wide range, sometimes feeding in the top of 30-meter trees and at other times perching at eye-level. They fly with whirling wing-beats, maneuvering about with lightning speed, constantly keeping in contact with clicking, metallic "ticks."

Oriental White-eye

Zosterops palpebrosus

A pretty and active warbler-like bird, the Oriental White-eye is widespread throughout Southeast Asia and can be very common locally. This bird is unusual in several ways. It is one of the few species found in primary rainforest that can adapt to a variety of habitats including mangroves, coastal woodlands and disturbed scrub. Unlike most other forest birds (which are usually solitary or seen in pairs) the Oriental White-eye travels in flocks, often high at the very top of the forest canopy. A nimble and energetic bird, it moves restlessly through the foliage, on the lookout for small insects, berries and nectar, constantly calling out with a faint, ringing voice. It bathes in the small pools of water that collect on branches in the treetops.

The white-eye is easy to keep in captivity and is a fairly popular cage bird. Several very similar species occur at montane elevations and are endemic to some of the Southeast Asian islands. Where their distributions overlap, identification can be extremely difficult.

Family:
Zosteropidae

Distribution:
South Asia;
Indonesia,
Malaysia,
Thailand and
Singapore.

Index of Common Names